The Compound of Five Cubes

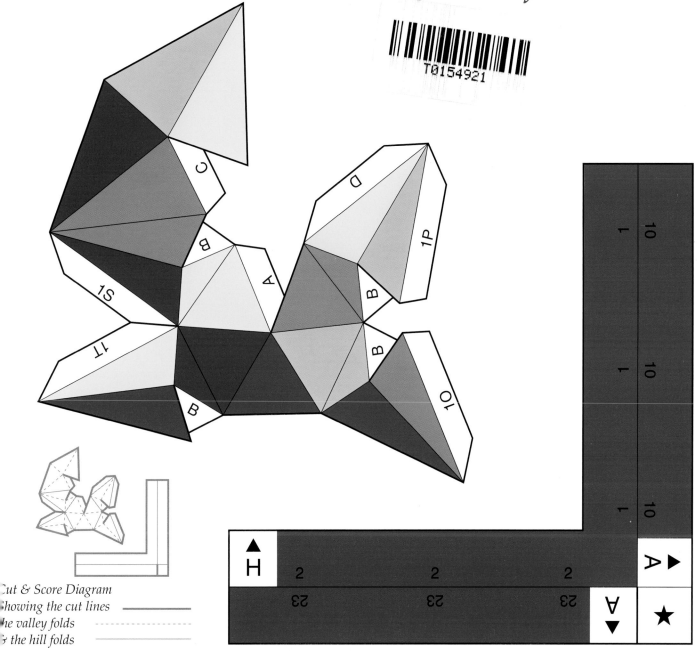

*Cut & Score Diagram
showing the cut lines* ——————
the valley folds - - - - - - - -
& the hill folds ——————

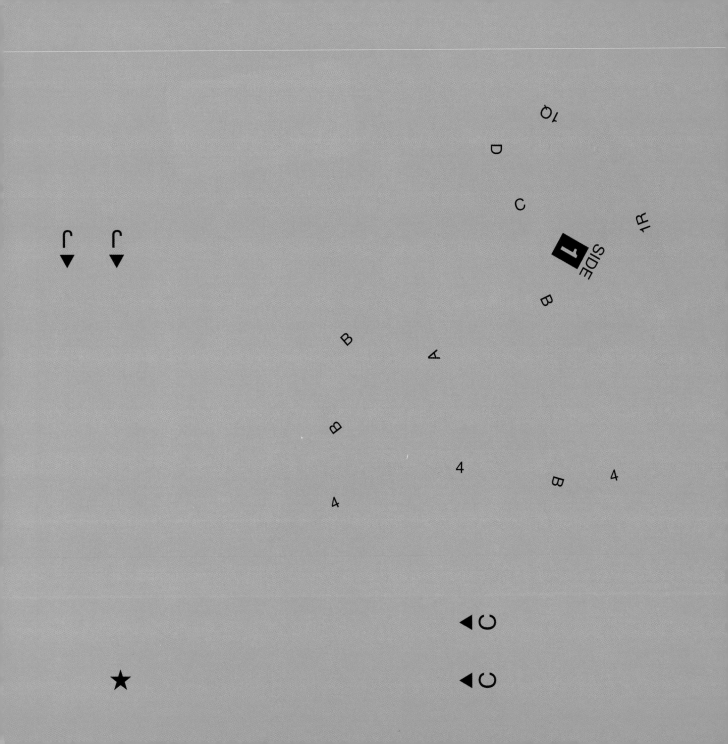

The Compound of Five Cubes

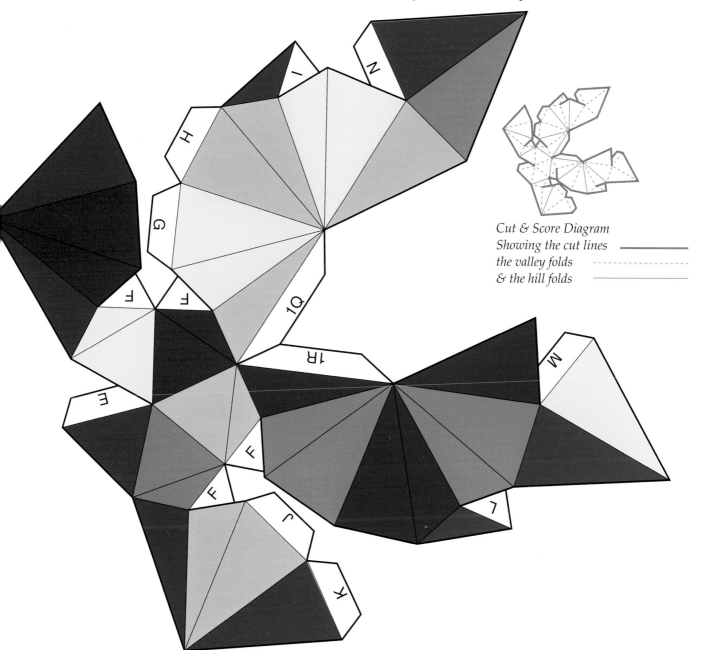

Cut & Score Diagram
Showing the cut lines
the valley folds
& the hill folds

3

1O   N   I   3

H
3

1P

G

F   F   2

E

M   1S

1T   F   2

1   L   J

1   K   F

L SIDE   2

1

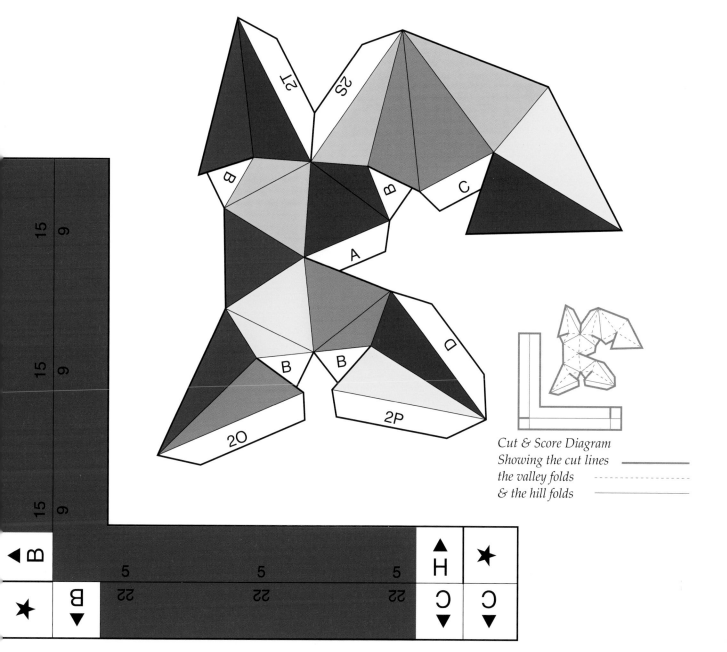

Cut & Score Diagram
Showing the cut lines
the valley folds
& the hill folds

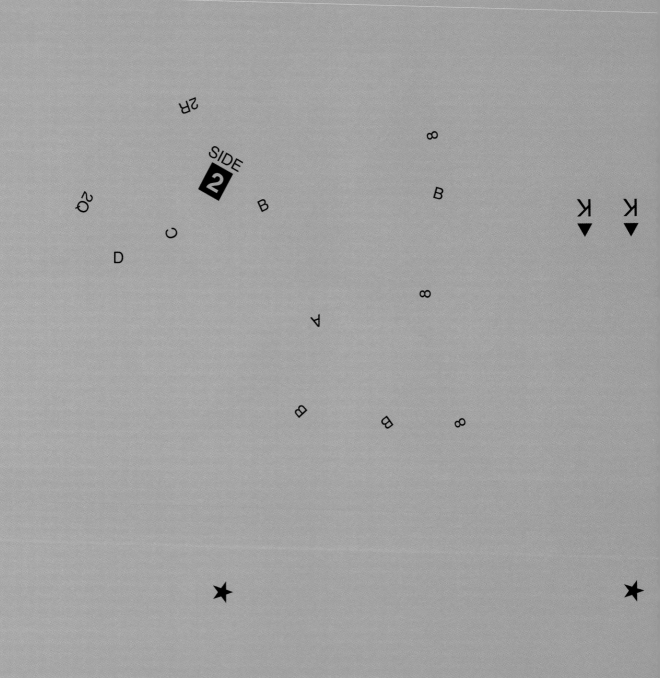

The Compound of Five Cubes

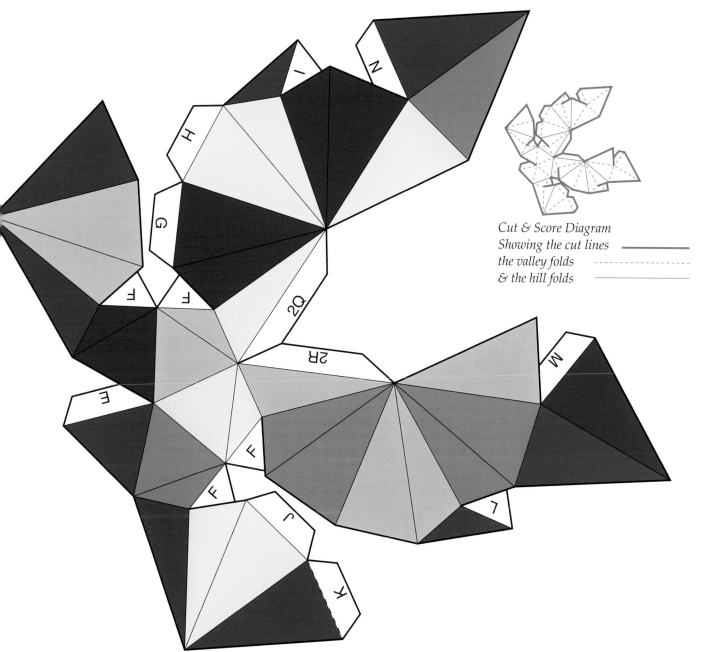

Cut & Score Diagram
Showing the cut lines
the valley folds
& the hill folds

L

2O

N I L

H

L

2P

G

F

F

6

E

M 2S

2T

F

5

L J

6

5 K

F

5 SIDE 2

6

5

The Compound of Five Cubes

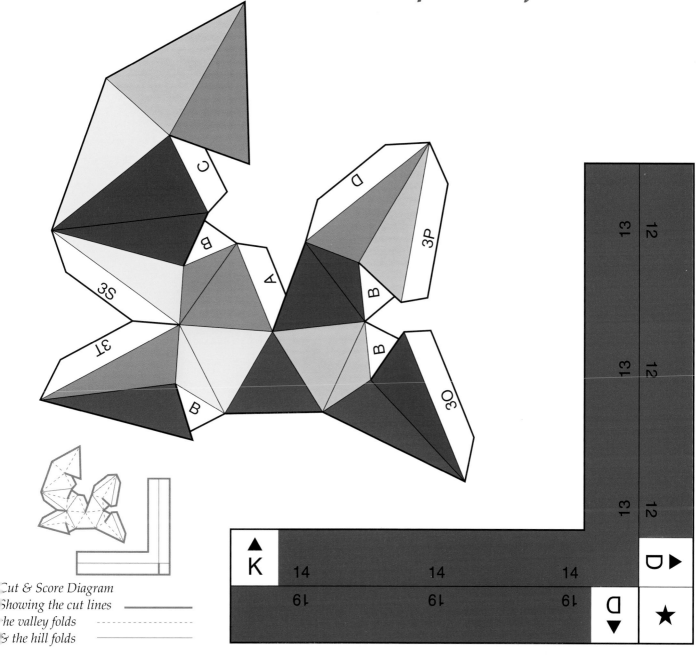

Cut & Score Diagram
Showing the cut lines ———
the valley folds ----------
& the hill folds ———

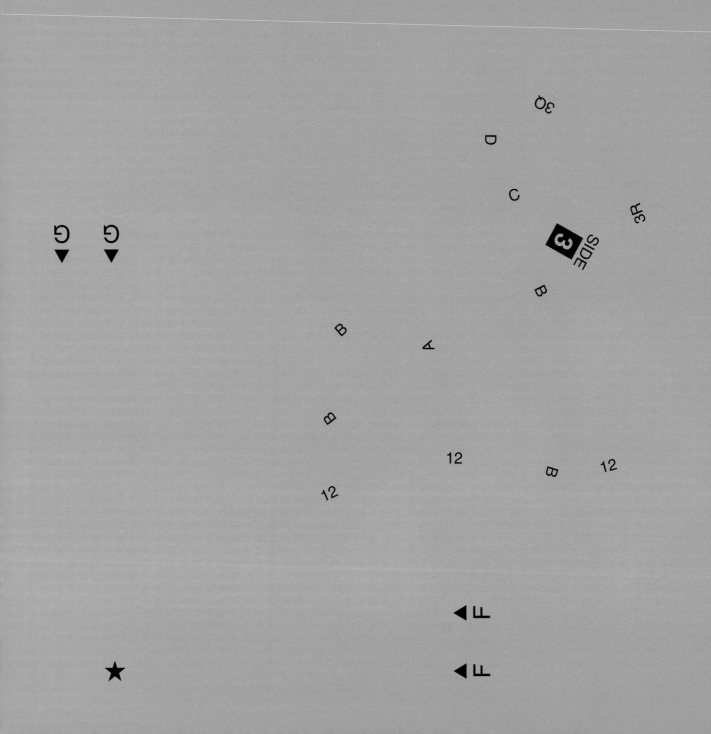

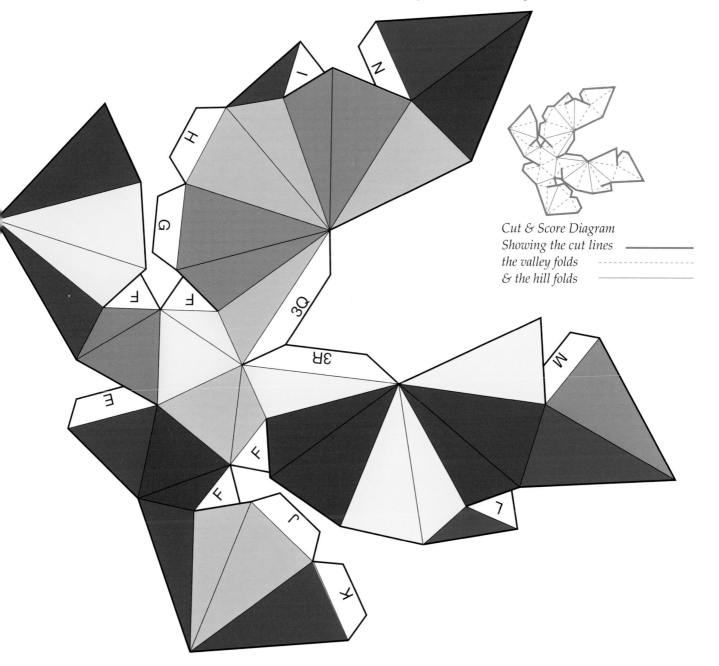

Cut & Score Diagram
Showing the cut lines ————
the valley folds - - - - -
& the hill folds ————

11

3O

N I 11

H 11

3P

G

F F 10

M 3S

E

3T 10

9 F

L J

9 K F

SIDE ◆3 10

9

The Compound of Five Cubes

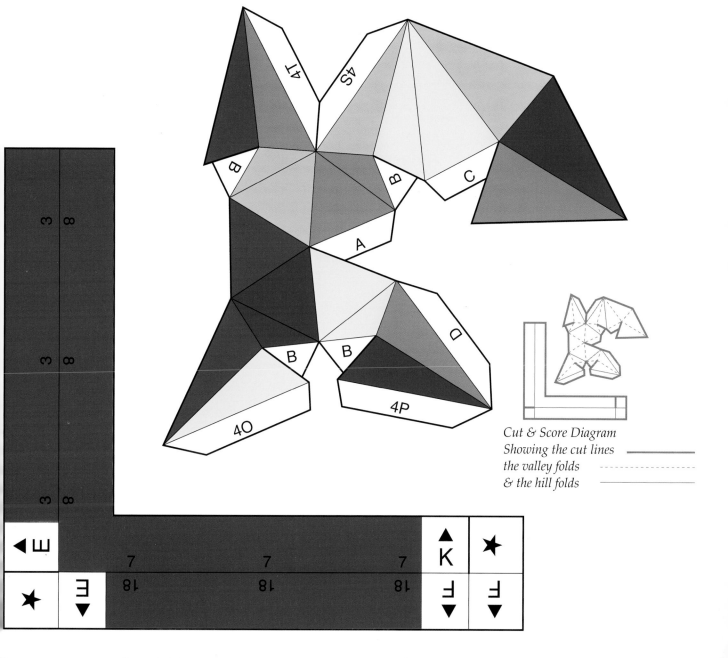

Cut & Score Diagram
Showing the cut lines
the valley folds
& the hill folds

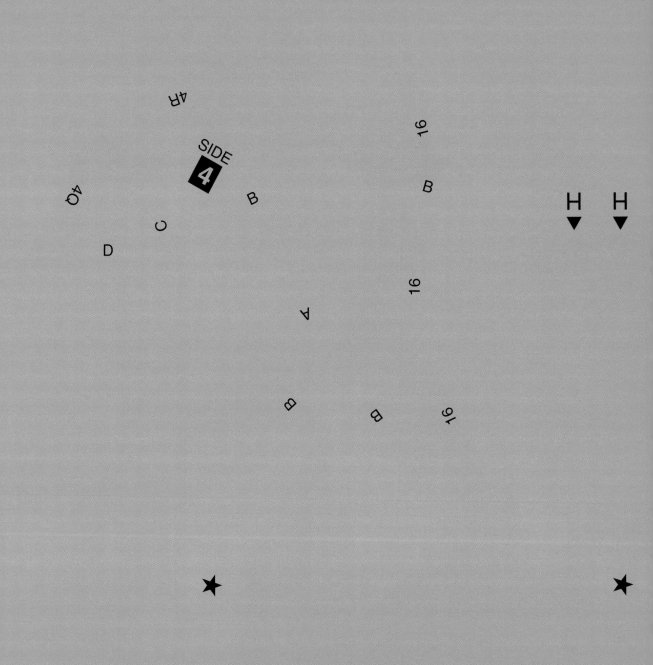

The Compound of Five Cubes

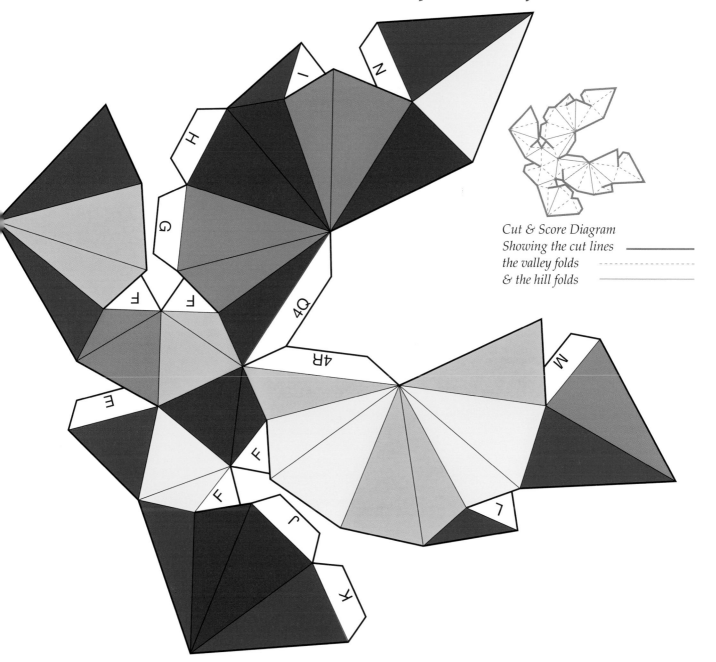

Cut & Score Diagram
Showing the cut lines
the valley folds
& the hill folds

15

4O

N I 15

H

15

4P

F F

14

G

M 4S

E

4T

14

13

F

L

J

13 K F

SIDE 4

14

13

The Compound of Five Cubes

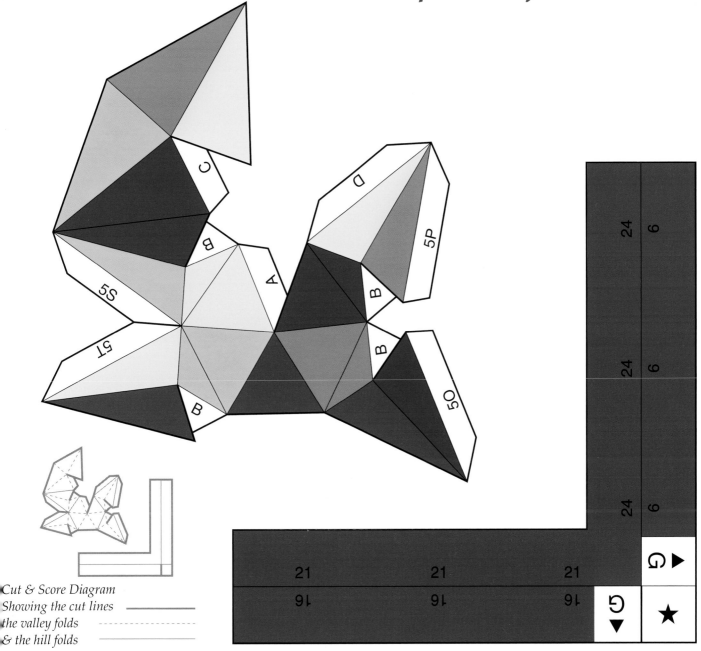

Cut & Score Diagram
Showing the cut lines ———
the valley folds ------------
& the hill folds ———

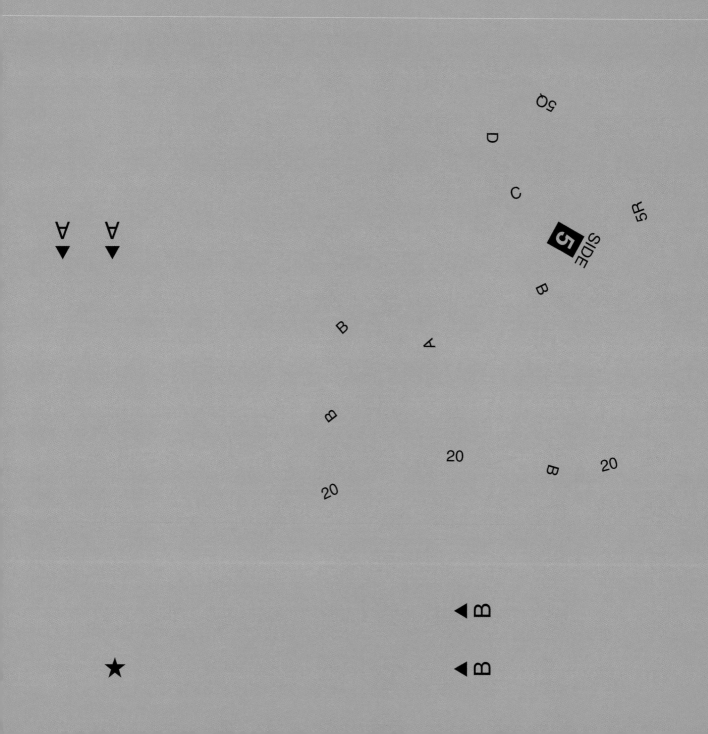

The Compound of Five Cubes

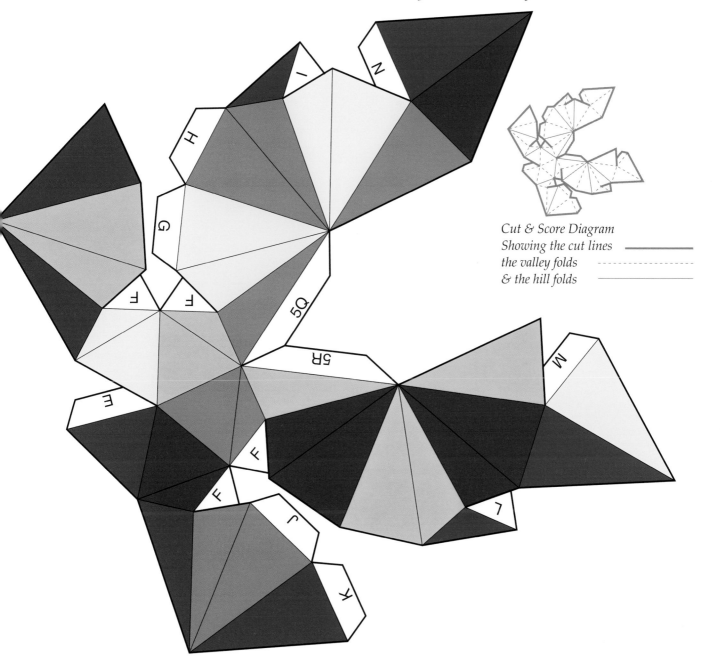

Cut & Score Diagram
Showing the cut lines —————
the valley folds - - - - - -
& the hill folds ———————

19

5O

N I 19

H

19

5P

G

F F 18

E

M 5S

5T

F

18

17

L J

17 K F

18

SIDE 5

17

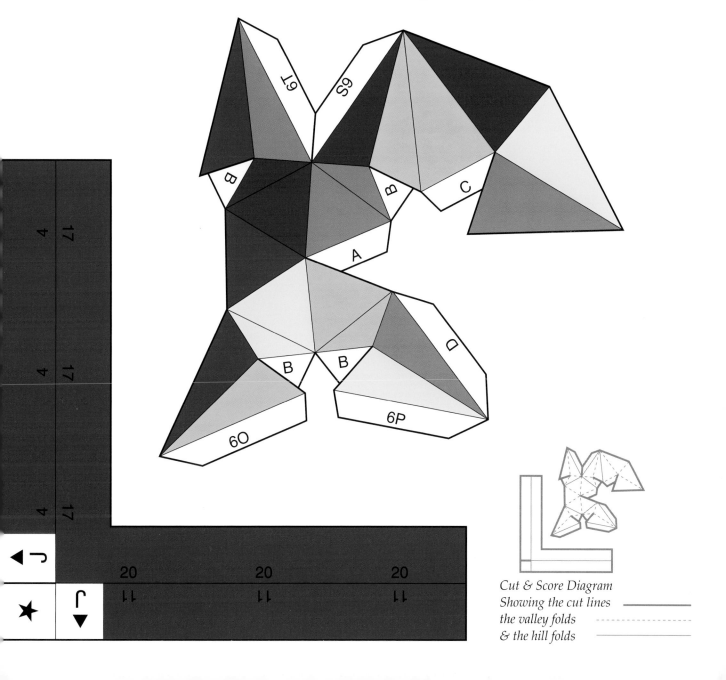

Cut & Score Diagram
Showing the cut lines ———
the valley folds - - - - -
& the hill folds ———

The Compound of Five Cubes

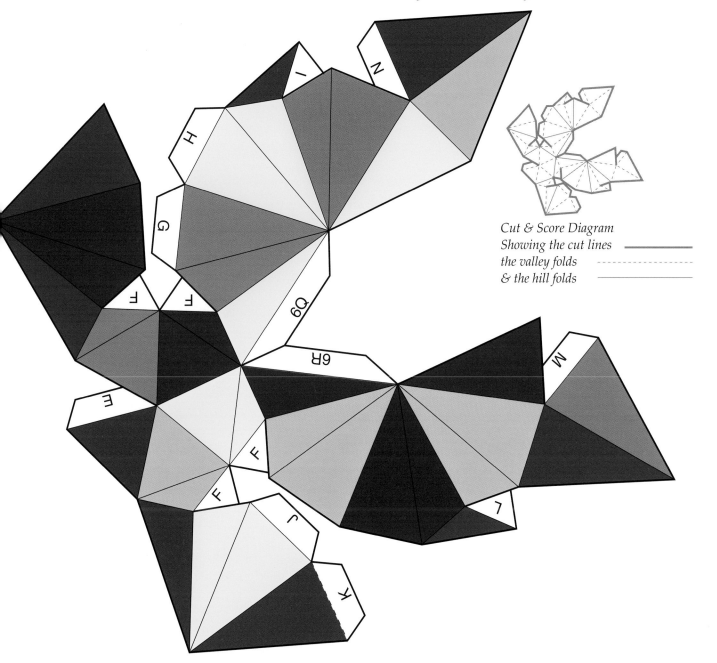

Cut & Score Diagram
Showing the cut lines ————
the valley folds - - - - -
& the hill folds ————

23

O9

N I 23

23

H

P9

G

F F 22

E

M 4S

22

T9

21

F

L J

21 K

F

SIDE 9

22

21